SCIENCE

Microscope Activities

HERON BOOKS
K-12 CURRICULUM

Additional Resources

For a free learning guide, go to *heronbooks.com/learningguides*.

For exam, email *teacherresources@heronbooks.com*.

We would love to hear from you!
Email us at *feedback@heronbooks.com*.

Published by
Heron Books, Inc.
20950 SW Rock Creek Road
Sheridan, OR 97378

heronbooks.com

First Edition © 1983, 2019, Heron Books
All Rights Reserved

ISBN: 0-89739-119-5

Printed in the USA

10 February 2019

CONTENTS

1 THE LIGHT MICROSCOPE 5

    Lenses 6

2 PARTS OF THE MICROSCOPE 9

3 USING THE MICROSCOPE 13

    Focus 13

    Controlling the Light 14

    Slides 15

4 PRACTICE USING A COMPOUND MICROSCOPE 17

    Activity #1 17

5 KEEPING YOUR MICROSCOPE CLEAN AND WORKING 21

6 WET MOUNT SLIDES 23

    Introduction 23

    Preparing and Disposing of a Wet Mount Slide 24

7 PRACTICE CREATING SLIDES 27

    Activity #2 27

8 WET MOUNT SIMPLE STAIN 29

9 CELLS UNDER THE MICROSCOPE.....................................31

Onion Cells..31

Human Cells..33

10 BEGINNING THE STUDY OF MICROSCOPIC LIFE...............35

Bacteria...35

Algae and Flagellates...35

Amoebas..37

Ciliates...37

Rotifers..37

Worms..37

Tiny Crustaceans 38

Fungi..38

11 OBSERVING BACTERIA UNDER A MICROSCOPE.................39

Activity #3 39

12 HANGING DROP MOUNTS...41

Method 1 42

Method 2 43

The Light Microscope

1

A **microscope** is an instrument that uses lenses to greatly magnify things that are invisible or hard to see. A **light microscope** (also called an **optical microscope**) is a microscope that uses light to magnify tiny objects.

A magnifying glass is a single lens and could be thought of as a simple kind of light microscope. The curved glass of the lens spreads out the light and makes an object appear larger than it actually is.

The most common type of microscope is a **compound microscope**, a microscope that uses two or more lenses together to magnify an object.

LENSES

The most basic compound microscope is a tube with lenses at both ends. The lens nearest the object you are viewing is called the **objective lens**, usually simply called the **objective**. The objective magnifies the image of the object.

The **eyepiece lens** (usually just called the **eyepiece** or the **ocular**) then magnifies the image again so that the final image may be hundreds of times larger than the original object.

The illustration on this page shows the relationship of the lenses, object and light source.

The amount that an image is enlarged by a lens is called the power of the lens, or its **magnification**. A lens which magnifies objects to 10 times their original size would have a magnification of "10x" (10 times).

A student microscope typically has three objectives of different powers. The magnification of each lens is shown somewhere on the metal cylinder that holds the lens and can be easily found. An example is shown on the next page.

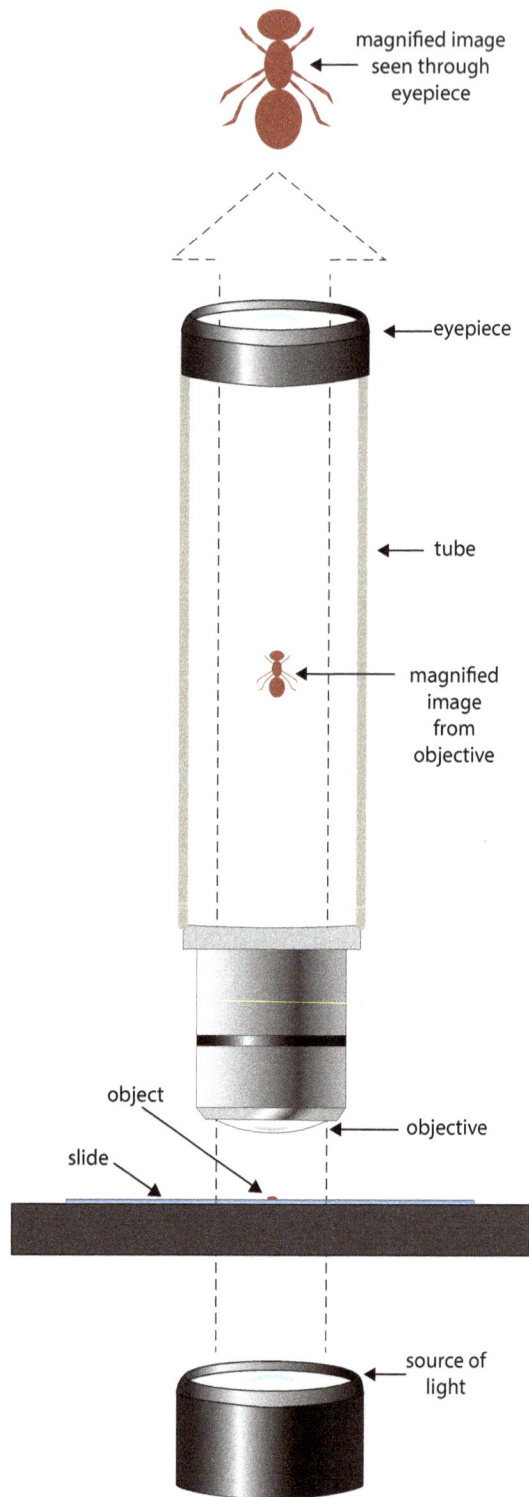

magnified image seen through eyepiece

eyepiece

tube

magnified image from objective

object

slide

objective

source of light

The total magnification of a microscope is computed by multiplying the magnification of the objective times the magnification of the eyepiece. The eyepiece is usually 10x. The low-power objective might be 4x, the medium-power objective 10x and the high-power objective 40x. The final magnification for each of these objectives then is 10 x 4 = 40x (low power), 10 x 10 = 100x (medium power) and 10 x 40 = 400x (high power).

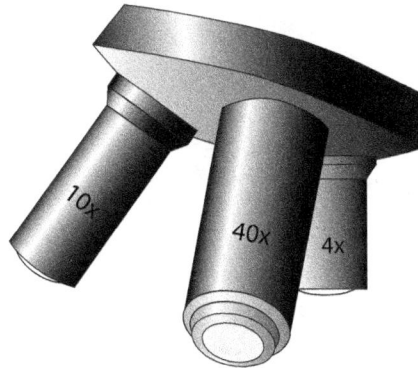

Higher quality microscopes may have higher-power objectives, including a 100x objective (total magnification 1000x) for looking at bacteria and tiny particles.

The ability of a lens or a microscope to produce clear and sharp detail is called **resolution**. As the magnification is increased, it is easier to see details more clearly and sharply. But resolution might not always increase as much as the magnification does because other things can make the image less clear. For instance, the lenses may not be totally smooth and so distort the light, or the lighting may not be concentrated enough to see objects under the microscope well.

For these reasons magnification beyond about 500x is not very effective unless the lenses are very high quality. The limit of very good light microscopes is about 1000x.

Parts of the Microscope

2

To use a microscope easily and well, one should learn the names of its parts and the jobs that they do. The illustration of a student microscope on the next page gives this information.

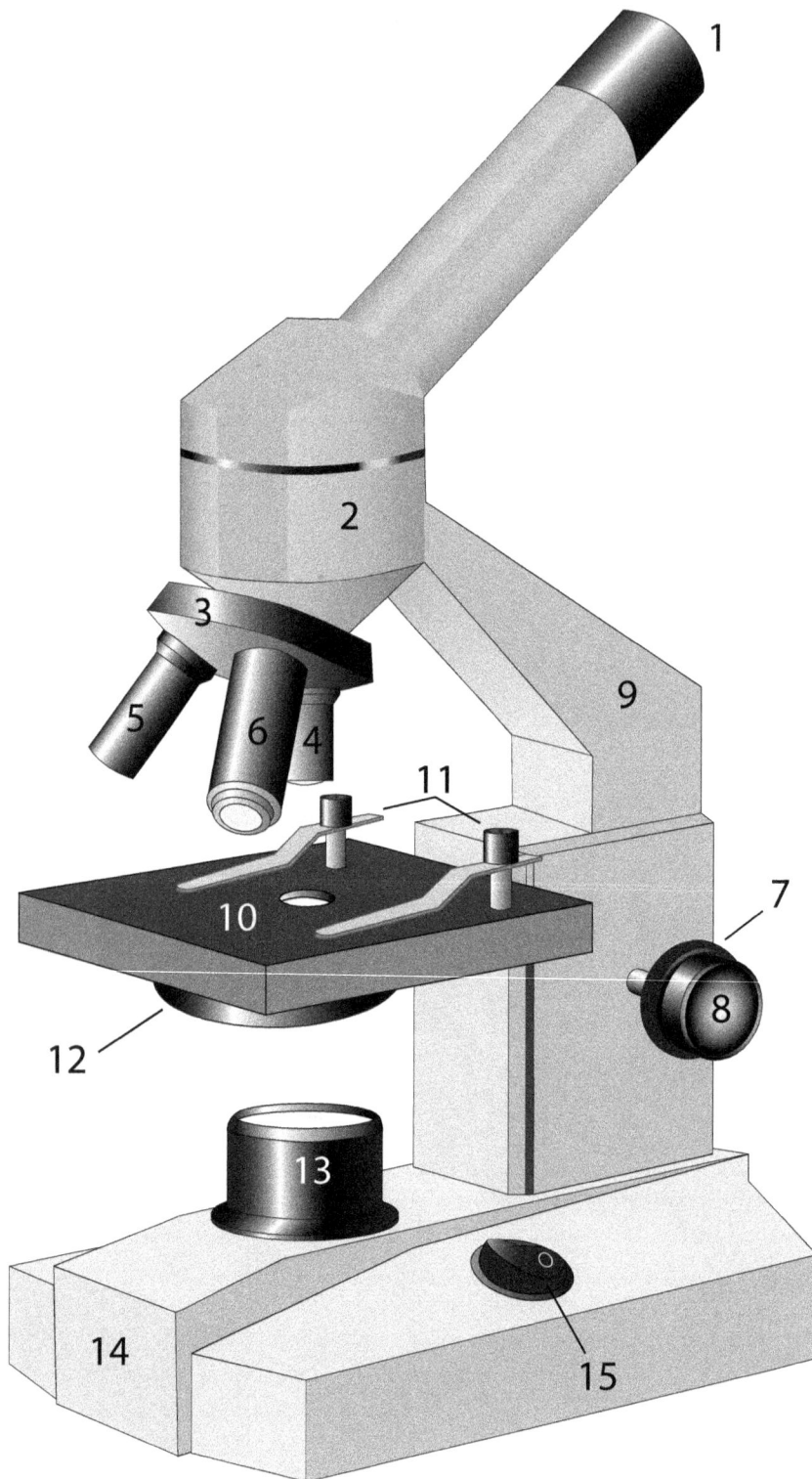

	PART	WHAT IT DOES
1	eyepiece	holds the lens the observer looks through, usually 10x
2	body tube	holds the eyepiece and objective at the correct distance apart
3	rotating nosepiece	holds the objectives; is rotated to change objectives
4	low-power objective	lowest power of magnification, usually 4x
5	medium-power objective	middle power of magnification, usually 10x
6	high-power objective	highest power of magnification, usually 40x
7	coarse adjustment (focusing) knob	moves the body tube (or stage) up and down in large movements
8	fine adjustment (focusing) knob	moves the body tube (or stage) up and down in smaller movements
9	arm	supports the body tube; handle for carrying microscope
10	stage	platform to support microscope slide or object being observed. There is a hole in the center so that light from below can strike an object placed on the stage.
11	stage clips	hold microscope slide in place
12	diaphragm (not on all microscopes)	controls the amount of light passing through the objective lens into the body tube
13	illuminator	directs light to the object and lenses
14	base	supports and stabilizes microscope, usually heavy
15	on-off switch	turns illuminator (light) on and off

Using the Microscope

3

FOCUS

Focusing is an important part of seeing the greatest detail. For good focus, an object needs to be just the right distance from the objective. Student microscopes usually have two adjustment knobs for focusing. The **coarse adjustment** knob brings the object into "preliminary focus" by moving the upper part of the microscope up and down in large movements. (Some microscopes are designed so that the **stage** moves up and down instead.) Then the **fine adjustment** knob brings the microscope into sharp focus by moving the microscope in very small movements.

At higher powers the fine adjustment knob is important because the coarse adjustment's movements are too big to focus on tiny cells with precision or are too big to focus at precise points of larger objects.

The up-and-down space, or vertical space, that is in focus is called the **depth of field**. As magnification increases, the depth of field becomes thinner or shallower (less deep).

10 x magnification   100 x magnification

Thus at high power the microscope has to be focused up and down with the fine adjustment knob in order to see the detail of a thick sample. If you are using high power and want to track a tiny organism that is swimming up and down, you may have to *continuously* focus the microscope with the fine adjustment knob in order to follow it.

CONTROLLING THE LIGHT

In order to see an object clearly, the microscope needs a good light source.

The light source must be in a position where it can be reflected up into the microscope. Most microscopes now have built-in lighting, sometimes called an **illuminator**.

Some microscopes have a device called a **diaphragm**—an adjustable disk mounted right under the stage. It is used to control the amount of light entering the microscope by controlling the size of the opening or **aperture** through which light passes. If a larger diaphragm opening is used than is needed, the light just tends to scatter and cause glare, making objects more difficult to see clearly. The circle of light coming through the diaphragm should be just large enough to fill the microscope field of view, but no larger.

The diaphragm controls the **contrast**—some parts of the object should be lighter or darker than the others in order to see details. If too much light or too little light is used, the whole object can appear too light or too dark to see much. You often will have to work with some trade-off between the brightness and contrast; for example, you might have to give up some brightness and clarity to get some contrast. All of this can be controlled by adjusting the aperture.

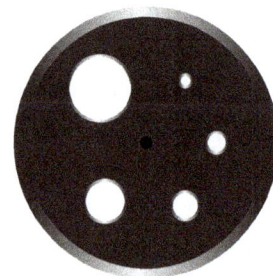

disc diaphragm

If you use a microscope with a diaphragm, practice will help you get good at this. Other microscopes use a dimmer switch to make the light not as bright.

SLIDES

Although not actually part of the microscope, slides should be mentioned here because they are normally used with a microscope. A **slide** is a very thin glass sheet in a rectangular shape that holds the sample to be viewed. Often an even thinner small sheet of glass or plastic called a **cover slip** is used on top of the slide with the sample between them. The slide sits on the stage and is held in place with **stage clips**.

Slides can be made by the person wanting to view specific things under the microscope, and instructions for doing this are given later in the book. Slides that are already made of plants, insects, animals, etc. (called "prepared slides") let you easily look at many other things that you might not be able to see otherwise.

Practice Using a Compound Microscope

4

These activities are to be repeated until you can do the steps with confidence and certainty. A microscope is required that has at least low- and high-power objectives (and preferably a medium-power objective), and coarse and fine adjustment knobs. Instructions for using the diaphragm need to be done only if your microscope has a diaphragm. The microscope must have a built-in light source.

ACTIVITY #1

1A. Set Up

1. Place the microscope on a table far enough away from the edge of the table that the microscope won't be tipped over.

2. If the stage, body or base of the microscope is dirty, wipe it with a piece of cheesecloth or other soft cloth, but don't wipe the lenses with a cloth.

3. If the microscope has an illuminator, wipe off the glass covering the top of the illuminator with moistened lens paper (moistened with water or lens cleaner (or fog the mirror with your breath) to avoid scratching it.

4. Clean the eyepiece lens and the objective lenses using lens paper as was done in the previous step. (Later, if you have trouble focusing, you can always check the lenses to make sure they are clean.)

5. Rotate the nosepiece of the microscope until the low-power objective is in position directly over the opening on the stage.

6. Try to keep both eyes open while you look through the eyepiece with one eye. (Your eyes will tire faster if you try to keep one eye shut for long periods of time.) If you wear glasses, you can take them off or leave them on as you choose.

7. Place a prepared slide on the stage, under the stage clips. Center it over the hole in the stage.

8. Turn the illuminator on.

1B. Low Power

1. Coarse Adjustment: With the coarse adjustment knob, lower the low-power objective[1] as far as it will go without hitting the slide.

2. Move the slide so that the object is directly under the low-power objective lens. Look through the eyepiece and adjust the low-power objective until the object comes into view.

3. (If there is a diaphragm) Adjust the aperture so that it is just large enough to light the microscope field, but no larger.

4. Slowly turn the coarse adjustment knob up or down to achieve the best focus you can.

5. Fine Adjustment: Focus the microscope on the object with the fine adjustment knob. Turn the knob in both directions to determine which way brings it into sharp focus.

   If you can't focus on it clearly, try these things to correct the problem:

   a) Check the position of the object to make sure that it is still centered.

---

1 For microscopes where the stage moves instead of the objective, substitute "raise the stage" for "lower the low-power objective."

b) Check the position of the objective to make sure that it is fully rotated into position.

c) Recheck the light source (and diaphragm, if you have one) to make sure you are getting the best lighting.

d) Clean the objective and eyepiece lenses again.

1C. Exercises

1. Do these exercises after set-up for low power:

a) Draw a picture of what you see under the microscope and save it for a later step.

b) Find out what happens when you move the slide to the right and to the left.

c) Find out what happens when you move the slide away from you and toward you.

d) Determine the total magnification under low power. The power of magnification (x) is written on the eyepiece and the objective. To determine the total magnification, multiply the number on the eyepiece times the number on the objective. Common objective lenses have a magnification of 4, 10, and 40. (Write your calculations on another sheet of paper.)

2. (If your microscope has a medium-power objective) Do these extra steps:

a) Make sure the object is still in focus under low power. While watching from the side, rotate the nosepiece until the medium-power objective is in position directly over the opening in the stage. Be careful that objective doesn't hit the slide. If the object is still in sharp focus, you can skip step b). If the object is now out of focus, do step b).

b) Adjustment: If the object is close to being in focus, bring the object to sharp focus with the fine adjustment knob. If the object is badly out of focus, first try to focus with the fine adjustment knob. If that doesn't work, carefully and slowly re-adjust the focus with the coarse adjustment knob while watching through the eyepiece. Then bring the object to sharp focus with the fine adjustment knob.

1D. High Power

1. Using low power or medium power, move the same prepared slide used above to the exact center of the field of the microscope and focus carefully. (The reason for moving to the exact center is so that it will still be in the field of view when you switch to high power.)

2. Move the high-power objective into position while watching from the side to make sure the high-power objective doesn't hit the slide. *At high power use only the fine adjustment knob to bring the object into sharp focus.* The high-power objective is so close to the slide that adjusting the coarse adjustment knob usually causes the objective to move too much, and could jam the objective into the slide. Doing so could dirty or scratch the objective lens, or break a slide or cover slip.

3. (If there is a diaphragm) Check the adjustment of the diaphragm to make sure that the circle of light is just large enough to fill the microscope field, but no larger.

4. Do the following steps on a separate sheet of paper:

    a) Draw a sketch of what you see.

    b) Compare this sketch to the picture drawn in 1C, step 1.a) above. What detail can you see now that you couldn't see under low power?

    b) What is the total magnification under high power? Show your calculations.

    c) How many times was the magnification increased when you went from low power to high power?

    d) How was the size of the area you could see affected when you changed from low power to high power?

Keeping Your Microscope Clean and Working

5

It is important to keep the microscope lenses and stage as clean as possible. If anything gets on a lens, it is likely to blur the microscope image so that you can't see clearly. If the stage gets dirty, the lens is likely to get dirty or sticky. If the slide gets sticky, it will stick to the stage and you'll have trouble moving it easily when you try to focus. Also, if you get the stage sticky, you are likely to get your hands and lens dirty too.

A microscope is a precise instrument that needs to be taken care of so that the parts work together as they should. If a microscope is jolted hard or dropped, or a knob forced to turn, parts may be damaged or thrown out of adjustment so they no longer work together properly.

Here are some things you can do to keep the microscope clean and working:

1. Never touch the lenses with your fingers or anything dirty. If the lenses become dirty, wipe them gently with lens paper. If a lens does not clean easily, use optical lens cleaner wipes or a soft cloth.

2. Do not tilt the microscope when working with liquids (water, oil, etc.) because the liquid may spill onto the stage, or run under it and clog microscope parts or make the stage sticky.

3. Do not leave a slide on the microscope when you are not using it or when you have finished with it. Material from the slide can get on the lens or on the stage.

4. Keep the stage of the microscope clean and dry. If you spill anything on the stage, clean it up immediately, and dry it with a cloth or paper towel. (If needed, also check under the stage to make sure that spilled liquid did not collect there.) If you get oil on the stage, wipe it off with a cloth or paper towel moistened with an optical lens cleaner.

5. If you have to carry a microscope from one place to another, use both hands so that you don't drop it and so that the eyepiece doesn't fall out.

6. All movable parts of a microscope should work freely. If any movable parts jam or stick, do not force them. Instead, see your supervisor for assistance. If you force a knob to work when it is jammed, you may ruin it.

7. When you store a microscope, always leave it with the low-power objective in the focusing position. (If the stage is movable, move it all the way down.) That way it is impossible to accidentally force the objective lens down into the stage and scratch the lens.

8. If possible, keep the microscope covered and stored in its own space when not in use. That will keep it from getting dirty, hit, dropped or knocked about.

Wet Mount Slides 6

INTRODUCTION

Learning to make your own slides opens up the microscopic world all around you. You can view almost anything under a microscope as long as the object you are viewing lies flat.

To make an object lie flat, use the tiniest sample/specimen (or part) and place it on the middle of the slide with the tiniest amount of water. Then a very thin square of glass or plastic called a **cover slip** (or **cover glass**) is placed on top. The cover slip holds down the sample and flattens it out. These are shown on the next page.

A slide prepared this way is called a **wet mount** and it is a good method for preparing temporary slides. The wet mount is temporary because the heat from the light source will usually cause the slide to dry up after a few minutes. To keep the slide from drying up, add another tiny drop of liquid at the edge of the cover slip.

PREPARING AND DISPOSING OF A WET MOUNT SLIDE

1. Get a clean slide. If it has any water marks on it, wipe it off with a paper towel so that the surface is clean.

2. Lay the slide flat on a table.

3. Place a tiny sample or specimen of what you want to observe in the center of the slide. Then add a tiny drop of water on top of it with a dropper.

    or

    Place a tiny drop of water in the center and then add the sample/specimen.

    or

    If the sample is liquid, just place a tiny drop of it in the middle of the slide.

4. Get a clean cover slip. Using a pair of tweezers to hold the cover slip while you lower it, touch one edge to the drop and lower the cover slip to the slide. Lower it slowly so that all the air bubbles are pushed out as the weight of the cover slip spreads out the drop. (Try this with tweezers first. But, with care, you can hold the cover slip by the edges between a thumb and a finger instead of using tweezers.)

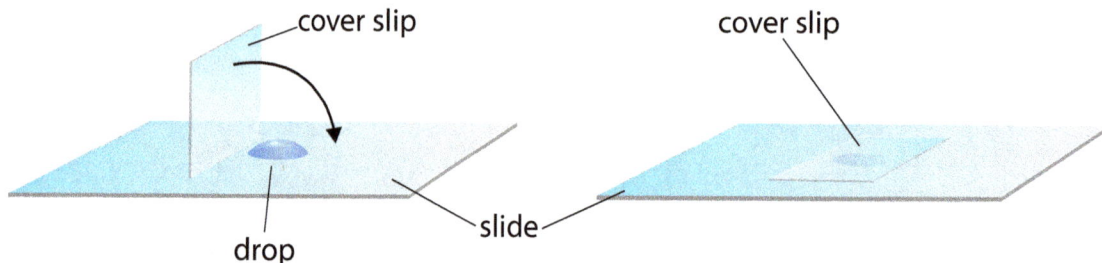

Note: It usually isn't a good idea to push down on a cover slip on a slide because doing so may leave fingerprints on the cover slip, break the cover slip, or push out the liquid and make a mess. The exception is when you need to squash something, such as a bug. In that case use very little or no water, put a piece of paper towel on top of the cover slip (to avoid a thumbprint) and push down gently.

5. The slide is now ready for observation unless the drop did not fill out evenly under the whole cover slip.

   a) If there is air space under the cover slip because there wasn't enough liquid, touch another drop of water to the cover slip edge and the water will be sucked under the cover slip and fill it out.

   b) If the drop was too large in the first place, liquid will overflow around the edges of the cover slip. In this case, touch the edge of a paper towel to the excess liquid until it is soaked up.

   c) Air bubbles are sometimes trapped under cover slips. They can be small or large, but they look more like round inner tubes than anything alive. They usually appear black around the edges and white or gray in the center. A few small air bubbles don't matter, but you may have to make a new slide if the air bubbles are large or numerous.

6. A container should be kept for disposing of used slides. It should be wide enough that you can easily drop in slides (or slides with cover slips) when you are finished. The container should be filled with enough disinfecting solution[2] to cover the slides. The solution is intended to kill any organisms that were in the wet mount. Later, the slides can be cleaned and reused.

---

2 A 10% bleach solution (10 parts bleach, 90 parts water) or a disinfecting cleaning solution will do.

Practice Creating Slides

7

ACTIVITY #2

2A. Low Power

1. Mount two different colored hairs, fibers or threads across each other on a clean slide. Make a wet mount by adding a drop of water where the two hairs cross and then covering the hairs with a cover slip.

2. Focus under low power. Move the slide so that the strands cross in the exact center of the field that you can see under the microscope.

3. With the fine adjustment knob try to focus on one strand and then the other. (Under low power both strands should remain in focus at the same time.)

2B. High Power

1. Move the strands to the center of the low-power or medium-power field. Then move the slide so that the strands cross directly in the center of the field.

2. Shift to the high-power objective and sharpen the focus with the fine adjustment knob *only*. Readjust the diaphragm as needed.

3. Bring the strands into sharp focus with the fine adjustment knob. Focus on the higher strand, then focus on the cross-over point of the strands, and then on the lower strand. Finally, return to the higher strand.

4. Answer on a separate sheet of paper: How can you tell which strand is on top?

Wet Mount
Simple Stain

8

After you have made a wet mount and looked at it under the microscope, you can improve the contrast by staining the wet mount in place. (Note: Some stains kill the organisms; blue food coloring shouldn't.)

1. Touch a small drop of blue food coloring to an edge of the cover slip. (Other stains such as methylene blue could be used as well, but see note above.)

2. Touch the edge of a paper towel to the opposite edge of the cover slip. This will pull out water on one side and will pull in the stain to replace it on the other side.

3. Observe your wet mount slide under the microscope while doing this and you will see the object getting darker as it is stained.

4. If the background turns dark just as fast as the specimen, you've put too large a drop of stain on the slide. Try removing the excess stain: a) add a drop of clear water to the edge of the cover slip, b) pull out the stain from the other side with a piece of paper towel. The specimen retains the stain.

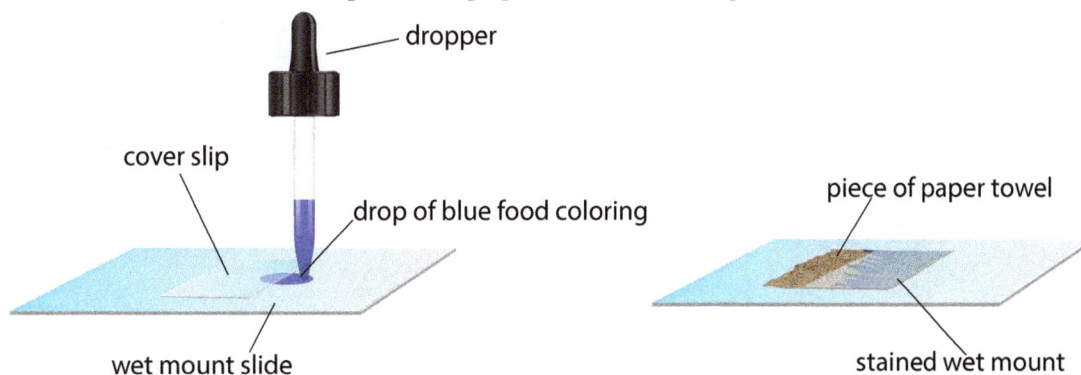

dropper

cover slip

drop of blue food coloring

wet mount slide

piece of paper towel

stained wet mount

If the background is still too dark, make a new wet mount using a smaller drop of stain. With the right amount of stain, the specimen will stain and turn dark much faster than the background.

Cells Under the Microscope 9

Materials needed for each activity: Optical microscope, light source, slides, cover slips, water, blue food coloring (or another stain), paper for sketching.

ONION CELLS

Additional materials needed: Onion, knife, tweezers, razor blade.

Onion cells are easy to prepare and observe under the microscope. Prepare a slide for observation of onion cells as follows:

1. Get an onion and slice it in half.

2. Peel off one of the layers.

3. On the inside of each layer is a very thin "skin." Find the skin and with your fingernails (or tweezers) carefully strip off a very thin piece of the skin. It should be clear.

4. Place a small piece of the onion skin (about ¼ inch by ¼ inch or smaller) on the center of a slide. If the piece is too large, carefully cut it to size on the slide with a razor blade.

5. Add a drop of water and make a wet mount with a cover slip.

6. Look at the slide under low- and high-power magnification. Compare what you see with the description and illustration below:

Under low power you can see layers of cells. Under high power, the cells will look larger and show more detail. Focus on a single layer of cells near the edge of the onion sample. The shape of the cells is rectangular. Inside the cell is a large circular "bump" which is barely visible unstained, but which will appear dark when stain is applied. This part of the cell is called the **nucleus**. The nucleus controls cell reproduction and heredity.

Surrounding the nucleus is the **cytoplasm**. It extends to the outer boundary, which is the cell wall. The cell wall gives the onion cell its rigid shape and doesn't stain at all. (Onion cells, like all plant cells, have cell walls.)

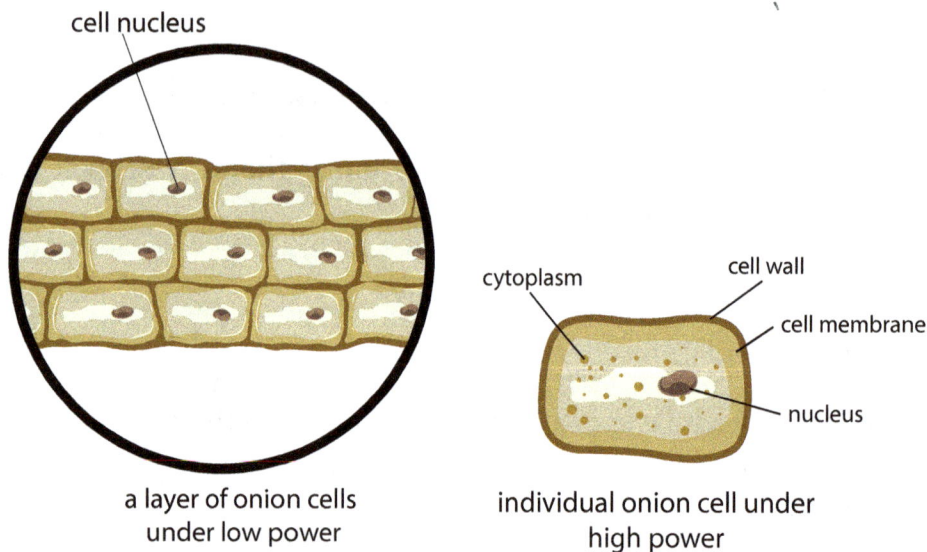

a layer of onion cells
under low power

individual onion cell under
high power

Both the nucleus and cytoplasm are living parts of the cell. Together they make up the **protoplasm**, the name given to all the living material found inside the cell. The entire protoplasm is "packaged" inside a thin membrane known as the cell membrane. The **cell membrane** is located just inside the cell wall and completely surrounds the cell and holds the contents of the cell together. The cell wall not only gives the cell a definite shape, but protects the membrane from splitting.

7. Stain the slide with blue food coloring (or another stain) and look at the slide again under low and high power. Again, compare what you see with the description.

   Note: If you use too large a piece of onion skin, it may not stain evenly. In that case, do this to even out the stain:

   a) Remove the cover slip, but leave the onion skin flat on the slide.

   b) Add a tiny drop of stain and enough water for the cover slip.

   c) Using a dry cover slip, make the wet mount again.

   d) Look at the stained slide again under low and high power.

8. Sketch and label what you see under low and high powers.

HUMAN CELLS

Additional materials needed: Toothpick

1. Place a drop of water on a clean slide.

2. Using a toothpick, gently scrape the inside of your cheek two or three times.

3. Make a wet mount—swish the end of the toothpick vigorously in the drop of water on the slide to help get the cells off the toothpick.

4. Put the cover slip on.

5. Look at the slide under low- and high-power magnification. Compare what you see with the illustration (which is about 10 times larger than what you will see) and description that follows:

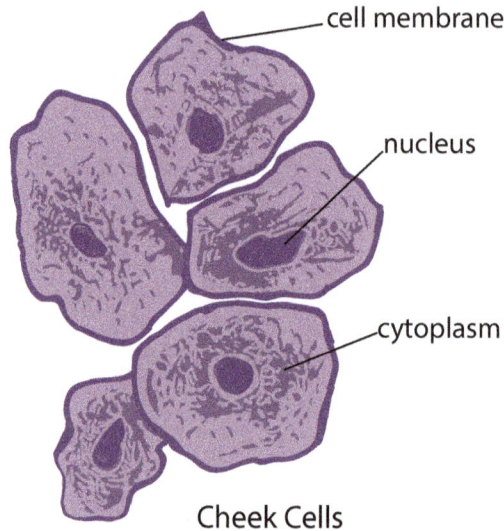

cell membrane

nucleus

cytoplasm

Cheek Cells

Human (and animal) cells have a nucleus, cytoplasm and a cell membrane, but they don't have cell walls to keep the cell rigid as plants do.

6. Stain the slide and look at the slide again under low and high power. Again, compare what you see with the description.

7. Sketch and label what you see under low and high powers.

Beginning the Study of Microscopic Life

Many different kinds of microscopic organisms can be seen in samples under the microscope. Here are some tips for recognizing and distinguishing some of the kinds of organisms you might find while exploring.

BACTERIA

Bacteria are tiny single-celled organisms that are so small they usually have to be magnified 400 to 1,000 times to be seen at all (unless they are growing in huge clusters in which case you might be able to see them without a microscope). They are usually much smaller than other kinds of organisms you will see under the microscope. Their main job is to decompose dead plants and animals so the nutrients from them can be recycled to grow new plants and animals. They also are the food for some larger microbes and tiny animals, like earthworms. Bacteria are found almost everywhere on the planet (including more than a thousand different kinds of "friendly" bacteria that live in our own intestines and help protect us from disease-causing bacteria).

ALGAE AND FLAGELLATES

Algae are plant-like organisms that contain **chlorophyll**, a green pigment that traps energy from sunlight. Both algae and green plants use energy from chlorophyll to live and grow new cells. Most algae grow as single cells but some algae cells form structures that look something like leaves or plant stalks, but they don't form true roots and leaves like plants. Most kinds of algae cells are microscopic, but like bacteria can be seen without a microscope if enough cells are clustered together. In fact, some large marine algae, called brown seaweed, can grow more than 100 feet long! Creatures ranging from microscopic size to whales eat algae, while other animals eat the tiny organisms that eat the algae.

Algae come in a beautiful variety of colors and shapes, and they are found in many places on land as well as in water. They all have a green coloring, but the green may be masked by other colors. The main microscopic forms of algae are bright green or golden brown, and sometimes blue-green (a darker green). Algae in fresh water, such as in ponds, lakes and ditches, sometimes form scum, and grow as green, hairy growths on objects under water. Bright green paint-like patches on rocks and trees often contain green algae. Darker green patches around wet places, such as water fountains and swimming pools, may be blue-green algae. In cleaner waters, single-celled yellow-green algae called **diatoms** often form golden brown coatings on rocks.

Flagellates are transparent one-celled organisms that pull themselves along with a "whip" that is attached to the body. Usually the whip, called a **flagellum** (which is Latin for "whip") is moving too fast for you to see, but you might be able to see it by adding a chemical that slows it down. These organisms usually live on waste materials as do bacteria. But there is an overlap between flagellates and algae. Some single-celled flagellates have a green pigment and are classified *either* as algae or flagellates.

A few types of flagellates group together and form balls of cells, called **colonies**, with all the flagella oriented outward, and sometimes they are big enough to see without a microscope. Again there are transparent forms and green forms, and the green colonial forms are called **colonial algae**.

AMOEBAS

If you see something under the microscope that looks like a tiny but dirty blob of jelly, watch it for a moment or two. It may extend part of its main body into something that looks like an arm or a leg and then draw it back again. If you see this happen you are looking at an amoeba (also spelled ameba). **Amoebas** crawl around to find and consume bacteria.

CILIATES

Ciliates are little animal-like creatures that have many **cilia** (hair-like projections on the surface). Even though they look complicated, each ciliate is just a single cell. Some types of ciliates have cilia over their entire surface. In other types the cilia occur in patches or are stuck together to form larger hairs. The cilia beat together in unison to help the creature swim or to bring in bacteria for food. Without a high-powered microscope, the hairs are very small and difficult to see.

ROTIFERS

Rotifers are sometimes mistaken for ciliates, but they are multicellular and have cilia only at one end. The cilia on the front part of the rotifer whirl and suck in water very powerfully. In fact, the cilia of rotifers beat so strongly that under the microscope you can see debris being stirred up and sucked in, as if the animals were vacuum cleaners. At the opposite end of the rotifer is a base or foot that the rotifer uses to attach to objects. Rotifers eat bacteria, algae or small ciliates.

WORMS

There are many different kinds of worms and all of them are multicellular. Most of them live in rich decaying soils or in masses of decaying algae. Some are flat, some are round, and some are segmented (like earthworms). Most of them are visible to the eye or with low power under the microscope.

TINY CRUSTACEANS

Crustaceans are a very large group of multicellular animals that are related to insects and spiders. They have many jointed legs and bodies similar to shrimp or crab (both of which are also in this group). Nearly all of them live in water and their size ranges from microscopic to lobster size. The tiny ones, sometimes called "micro-crustaceans," are eaten in great numbers by small fish and other tiny animals. The ones most likely to be found in a drop of water are barely visible to the naked eye or can be seen with low power under the microscope. Three common groups of tiny crustaceans are fairy shrimps[3], water fleas and copepods (including "Cyclops" that has one red eye).

FUNGI

Fungi are a group of organisms that can resemble algae, but are often colorless. They do not have chlorophyll, so for energy they absorb nutrients (rather than use sunlight). Fungi, like bacteria, help to make things decay, particularly the parts of plants most resistant to bacterial decay. Mushrooms are some of the largest and best known of the fungi, but molds found on stale bread and rotting fruit also are well known. Most fungi (like the ones mentioned) are multicellular, but some are single-celled, such as the yeast used to make bread and wine. Most fungi are visible to the eye, but to see the details of their structure, a microscope is needed.

---

3   This group includes the brine shrimp used by aquarium keepers as fish food.

Observing Bacteria Under a Microscope

11

ACTIVITY #3

3A. Grass

1. Pick a piece of grass and put it on a small dish of water.

2. Leave overnight.

3. The next day, make a wet mount slide of a small piece of the grass.

4. Look at the slide under low and high power.

5. Note any bacteria.

3B. Mouth Bacteria

1. Place a drop of water on a slide.

2. A good place to begin looking for bacteria is in your mouth. Get a toothpick and scrape off some of the white material (called "dental plaque") on your teeth (especially where the gum and teeth meet.

3. Place a tiny speck of this material in the water drop and mix it with the water using the toothpick.

4. Make a wet mount and look at the slide under low and high power. The material is so transparent that you will not see much detail. Most bacteria are too small to see under low power but some will be visible under high

power. The bacteria will be found mostly in clusters but some cells or small clusters of cells will be recognizable. When trying to focus on samples that are mostly transparent and can only be seen clearly under high power (such as unstained bacteria), it is helpful to purposely introduce a tiny air bubble. Then at a lower power, focus on the edge of the air bubble. That will help you find the depth where the bacteria are located.

5. Stain the wet mount and look at the slide again. Most of the bacteria will be found in large clusters of different shapes, but individual cells and small groups of cells will look something like this (but much smaller):

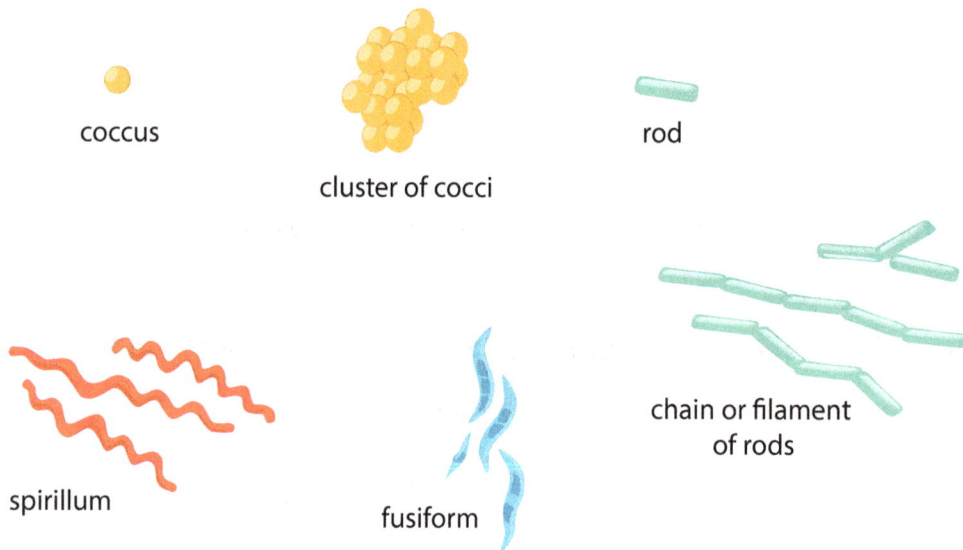

coccus

cluster of cocci

rod

spirillum

fusiform

chain or filament
of rods

Hanging Drop Mounts

For most microbes a flat slide mount is adequate, but sometimes microscopic organisms are too large to fit neatly under a cover slip on a flat slide, or you may want to give them more room so that you can watch their swimming motion. A different type of slide, called a **depression slide** or **well slide** can be used to accomplish this. It has a small depression in the center of one side.

Using this type of slide, the specimen is placed in a tiny drop on a cover slip, and the cover slip is then placed upside-down over the depression on the slide.

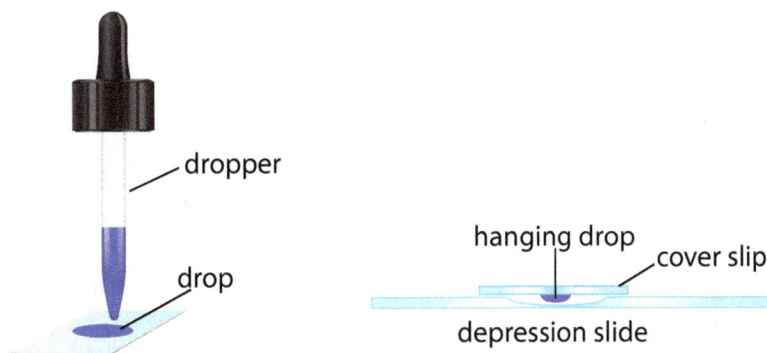

dropper

drop

hanging drop

cover slip

depression slide

METHOD 1

Making a Hanging Drop Mount Without a Depression Slide

If you don't have a depression slide, you can make a hanging drop slide using two regular slides, pieces of two flat toothpicks, a dropper and a drop of the liquid sample. This is an inexpensive way of making a hanging drop mount without a depression slide.

1. Break off the flat ends of two toothpicks. Then place one on top of each end of the lower slide (see illustration for placement).

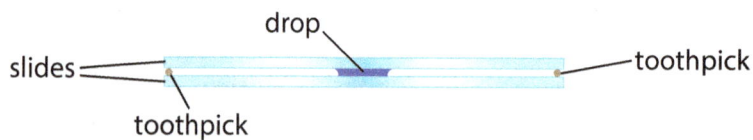

2. Using a dropper, place a *large* drop of liquid containing organisms in the middle of the slide.

3. Carefully lower a second slide onto the toothpick supports. The drop should touch the upper slide. If it doesn't, you need to reduce the distance between the slides or make the drop larger. (Remove the top slide, make sure the toothpicks are flat and/or increase the size of the drop, and replace the top slide.)

4. When the drop is large enough to touch the upper slide, carefully move the hanging drop slide onto the stage and examine it under low power. First find the edge of the drop under low power, then move toward the center of the drop.

5. If you have a medium-power objective, view the slide under medium power as well, but *avoid using the high-power objective*. The reason for not using high power with this mount is that the slide is too thick to focus through and the drop is too deep. If you attempt to use high power, you may break the slide or damage the objective.

METHOD 2

Making a Hanging Drop Mount Using a Depression Slide

1. Using a dropper, place a very *tiny* drop of liquid containing organisms in the middle of a clean cover slip.

2. Quickly invert the cover slip with the drop on it so that the drop now hangs down, and center the cover slip over the depression on the slide. If you invert the cover slip quickly, the drop of liquid won't run off.

3. Lower the cover slip gently to the slide.

4. Place the hanging drop slide on the stage and examine it under low power. To do this, first find the edge of the drop under low power, then move toward the center of the drop. If you have a medium-power objective, view the slide under medium power as well.

5. The focus in a hanging drop will be lower than with a regular wet mount because the drop hangs lower and deeper than it would on a wet mount slide.

6. If you want to look at the hanging mount under high power, switch to the high-power objective and refocus on the edge of the drop, then carefully move inward toward the center of the drop. Under high power, it is especially important that the drop be small and shallow. If the drop is *too big* and you try to focus at the bottom under high power, the objective may have to be focused downward so much that it breaks the cover slip.

    Note: Hanging drop slides, like wet mounts, tend to dry up. They can be made to last longer, if needed, by rubbing a thin ring of petroleum jelly around the edge of the depression before putting on the cover slip. This forms a tight seal and keeps the moisture from evaporating.

www.ingramcontent.com/pod-product-compliance
Lightning Source LLC
Chambersburg PA
CBHW051342200326
41521CB00015B/2595